HOW TO USE A METAL LATHE

Mastering the Art of Machining

Jeremiah Sand

ALL RIGHTS ARE RESERVED

No part of this publication may be reproduced in any form or by any means, including photocopying, recording, or any other electronic or mechanical methods without the prior written permission of the publisher except in the case of brief quotations embodied in reviews and certain other noncommercial uses permitted by copyrights law.

Copyright © Jeremiah Sand, 2024.

TABLE OF CONTENTS

CHAPTER 1 7
INTRODUCTION TO METAL LATHES 7
   1.1 What is a Metal Lathe? 7
   1.2 Types of Metal Lathes 8
   1.3 Anatomy of a Metal Lathe 9
   1.4 Basic Lathe Operations 10
   1.5 Safety Precautions for Lathe Use 10
   1.6 Setting Up Your Workspace 11
   1.7 Essential Tools and Accessories 12
CHAPTER 2 13
LATHE FUNDAMENTALS 13
   2.1 Understanding Cutting Tools 13
   2.2 Workholding and Chucking 16
   2.3 Measuring and Marking Techniques 17
   2.4 Cutting Speeds and Feeds 18
   2.5 Lubrication and Coolants 20
   2.6 Troubleshooting Common Problems 21
   2.7 Maintenance and Cleaning 22
CHAPTER 3 25
TURNING OPERATIONS 25
   3.1 Straight Turning 25
   3.2 Taper Turning 26
   3.3 Facing 27

3.4 Parting Off 28
3.5 Knurling 29
3.6 Thread Cutting 30
3.7 Advanced Turning Techniques 31
CHAPTER 4 33
DRILLING AND BORING OPERATIONS 33
4.1 Center Drilling 33
4.2 Drilling 34
4.3 Reaming 35
4.4 Boring 36
4.5 Counterboring and Countersinking 37
4.6 Thread Tapping 38
4.7 Honing 39
CHAPTER 5 41
THREADING OPERATIONS 41
5.1 External Threading 41
5.2 Internal Threading 43
5.3 Thread Repair 45
5.4 Choosing the Right Taps and Dies 46
5.5 Thread Measurement and Inspection 46
5.6 Threading Accessories 47
5.7 Troubleshooting Threading Issues 47
CHAPTER 6 49
FACING AND PARTING OPERATIONS 49
6.1 Facing Techniques 49

4

- 6.2 Parting Off 51
- 6.3 Grooving 52
- 6.4 Form Cutting 53
- 6.5 Cutting Off Bar Stock 54
- 6.6 Facing and Parting Tools 55
- 6.7 Safety Considerations 56

CHAPTER 7 57
PROJECTS FOR BEGINNERS 57
- 7.1 Making a Simple Bushing 57
- 7.2 Turning a Pen Blank 58
- 7.3 Creating a Candlestick Holder 59
- 7.4 Machining a Bottle Opener 61
- 7.5 Building a Small Engine Piston 62
- 7.6 Project Ideas for Inspiration 63
- 7.7 Troubleshooting and Refinement 64

CHAPTER 8 65
INTERMEDIATE LATHE TECHNIQUES 65
- 8.1 Eccentric Turning 65
- 8.2 Taper Attachment 66
- 8.3 Milling Attachments 67
- 8.4 Indexing and Dividing Head 68
- 8.5 Compound Rest Operations 69
- 8.6 Advanced Threading 70
- 8.7 Creative Lathe Projects 71

CHAPTER 9 72
ADVANCED LATHE OPERATIONS 72

- 9.1 CNC Lathe Programming................................ 72
- 9.2 Live Tooling 74
- 9.3 Multi-Axis Turning 75
- 9.4 Gear Cutting 76
- 9.5 Spline Cutting 77
- 9.6 Custom Tooling 78
- 9.7 High-Precision Machining 79

CHAPTER 10 81
MAINTENANCE, SAFETY, AND TROUBLESHOOTING 81

- 10.1 Regular Maintenance Tasks......................... 81
- 10.2 Troubleshooting Common Issues 83
- 10.3 Advanced Safety Protocols 85
- 10.4 Lathe Upgrades and Modifications............. 86
- 10.5 Resources for Further Learning 88
- 10.6 Building a Lathe Community 89
- 10.7 Sharing Your Lathe Creations 89

CHAPTER 1

INTRODUCTION TO METAL LATHES

1.1 What is a Metal Lathe?

A metal lathe is a powerful and versatile machine tool primarily used for shaping metal workpieces through a subtractive machining process. The workpiece is held and rotated by the lathe's spindle while a cutting tool is brought into contact with it to remove material, gradually shaping the workpiece into the desired form. Lathes are indispensable in various industries, including manufacturing, automotive, aerospace, and even hobbyist workshops.

The core principle of a lathe is relatively simple: rotation and cutting. The workpiece is secured to the spindle, which rotates at a controlled speed. The cutting tool, held in a tool post or turret, is then fed into the rotating workpiece, removing material in the form of chips. By controlling the movement of the cutting tool and the rotation speed, a wide range of shapes and features can be produced on the workpiece.

Key Advantages of Metal Lathes:

- **Versatility:** Lathes can perform a variety of operations, including turning, facing, boring, drilling, reaming, threading, and knurling.
- **Precision:** With skilled operation and proper tooling, lathes can achieve high levels of dimensional accuracy and surface finish.

- **Material Compatibility:** Lathes can work with a wide range of metals and alloys, including steel, aluminum, brass, copper, and titanium.
- **Customization:** Lathes can be equipped with various attachments and accessories to expand their capabilities and adapt to specific tasks.

1.2 Types of Metal Lathes

Metal lathes come in various types, each designed for specific applications and workpiece sizes. Here are some of the most common types:

- **Engine Lathe:** The most common type, offering a wide range of capabilities for general machining tasks.
- **Bench Lathe:** A smaller version of an engine lathe, ideal for hobbyists and small workshops.
- **Toolroom Lathe:** Designed for high precision and accuracy, often used for making tools and dies.
- **Turret Lathe:** Equipped with a turret that holds multiple tools, allowing for quick tool changes and efficient production of complex parts.
- **CNC Lathe:** Computer Numerical Control (CNC) lathes are automated machines that offer high precision, repeatability, and productivity for large-scale production.
- **Special Purpose Lathes:** These lathes are designed for specific tasks like crankshaft grinding, wheel turning, or gun drilling.

1.3 Anatomy of a Metal Lathe

Understanding the basic components of a metal lathe is crucial for safe and efficient operation. The main parts include:

- **Bed:** The foundation of the lathe, providing support and alignment for other components.
- **Headstock:** Located at the left end of the bed, it houses the spindle, gears, and motor that drive the workpiece rotation.
- **Spindle:** A hollow shaft that holds and rotates the workpiece.
- **Chuck:** A device attached to the spindle that grips the workpiece securely.
- **Tailstock:** Located at the right end of the bed, it supports the end of long workpieces or holds tools like drills and reamers.
- **Carriage:** A movable assembly that carries the tool post and cross slide, allowing for controlled movement of the cutting tool.
- **Cross Slide:** Mounted on the carriage, it moves the cutting tool perpendicular to the workpiece axis.
- **Compound Rest:** Mounted on the cross slide, it provides angular movement of the cutting tool for machining tapers and angles.
- **Tool Post:** Holds the cutting tool securely in place.
- **Apron:** Located on the front of the carriage, it houses the controls for feeding the cutting tool.
- **Leadscrew and Feedscrew:** Precision threaded rods that control the movement of the carriage and cross slide.

1.4 Basic Lathe Operations

Metal lathes can perform a wide range of operations, but some of the most fundamental ones include:

- **Turning:** Reducing the diameter of a workpiece by removing material.
- **Facing:** Creating a flat surface on the end of a workpiece.
- **Boring:** Enlarging an existing hole in a workpiece.
- **Drilling:** Creating a new hole in a workpiece.
- **Reaming:** Finishing a drilled hole to a precise size and smooth finish.
- **Threading:** Cutting threads on the outside (external) or inside (internal) of a workpiece.
- **Knurling:** Creating a textured pattern on the surface of a workpiece for grip or decoration.

1.5 Safety Precautions for Lathe Use

Operating a metal lathe can be dangerous if proper safety precautions are not followed. Here are some essential safety tips:

- **Wear Safety Gear:** Always wear safety glasses, hearing protection, and appropriate clothing.
- **Secure Workpieces:** Ensure workpieces are properly secured in the chuck or between centers.
- **Use Correct Tools:** Select the right cutting tools for the material and operation.
- **Avoid Loose Clothing:** Loose clothing, jewelry, or long hair can get caught in the rotating machinery.

- **Keep Hands Clear:** Keep hands and fingers away from the rotating workpiece and cutting tools.
- **Stop the Lathe:** Stop the lathe before making adjustments or measurements.
- **Never Leave Running Unattended:** Never leave a running lathe unattended.
- **Follow Manufacturer Instructions:** Read and follow the manufacturer's instructions and safety guidelines.

1.6 Setting Up Your Workspace

A well-organized workspace is essential for safe and efficient lathe operation. Here are some tips for setting up your workspace:

- **Adequate Space:** Ensure ample space around the lathe for movement and access to tools and accessories.
- **Lighting:** Provide good lighting to illuminate the work area and minimize shadows.
- **Tool Storage:** Organize tools and accessories in a designated storage area within easy reach.
- **Chip Removal:** Have a system in place for collecting and disposing of metal chips safely.
- **Emergency Stop:** Ensure the emergency stop button is easily accessible in case of an emergency.
- **Cleanliness:** Keep the work area clean and free of clutter to prevent accidents.

1.7 Essential Tools and Accessories

To get started with metal lathe operations, you'll need a set of essential tools and accessories:

- **Cutting Tools:** Turning tools, facing tools, boring bars, drills, reamers, taps, dies, etc.
- **Measuring Tools:** Calipers, micrometers, rulers, dial indicators, etc.
- **Workholding Devices:** Chucks, collets, centers, faceplates, etc.
- **Tool Holders:** Tool posts, turrets, quick-change tool holders, etc.
- **Lubricants and Coolants:** Cutting fluids, oils, greases, etc.
- **Safety Gear:** Safety glasses, hearing protection, gloves, etc.

By understanding the basics of metal lathes, their types, components, operations, and safety precautions, you'll be well on your way to mastering this versatile machine tool. As you gain experience and explore the various techniques and projects, you'll unlock the full potential of the lathe and create impressive metalwork pieces.

CHAPTER 2

LATHE FUNDAMENTALS

2.1 Understanding Cutting Tools

Cutting tools are the heart of any lathe operation, as they directly interact with the workpiece to remove material and create the desired shape. Understanding the different types of cutting tools, their geometries, and how to select and maintain them is essential for achieving precise and efficient machining results.

Types of Cutting Tools

- **Turning Tools:** Used for general turning operations, reducing the diameter of the workpiece.
- **Facing Tools:** Used for creating flat surfaces on the ends of the workpiece.
- **Boring Bars:** Used for enlarging existing holes or creating internal features.
- **Parting Tools:** Used for cutting off sections of the workpiece or creating grooves.
- **Threading Tools:** Used for cutting external or internal threads.
- **Knurling Tools:** Used for creating a textured pattern on the surface of the workpiece for grip or decoration.
- **Form Tools:** Used for creating complex shapes and profiles on the workpiece.

Tool Geometries

Cutting tool geometries play a crucial role in chip formation, cutting forces, and tool life. Some of the key geometric parameters include:

- **Rake Angle:** The angle between the tool's cutting face and a line perpendicular to the workpiece. A positive rake angle helps in chip flow and reduces cutting forces.
- **Clearance Angle:** The angle between the tool's flank and the machined surface. It prevents the tool flank from rubbing against the workpiece.
- **Nose Radius:** The radius of the tool's nose, affecting chip formation and surface finish.
- **Cutting Edge Angle:** The angle between the tool's cutting edge and a line parallel to the workpiece axis. It affects the strength and sharpness of the cutting edge.

Tool Materials

Cutting tools are made from various materials, each with its own properties and applications. Some common tool materials include:

- **High-Speed Steel (HSS):** A versatile and relatively inexpensive tool material suitable for a wide range of machining operations.
- **Carbide:** A harder and more wear-resistant material than HSS, ideal for high-speed machining and difficult-to-machine materials.
- **Ceramics:** Extremely hard and heat-resistant materials, used for high-speed machining of hard metals.

- **Coated Carbides:** Carbide tools with a thin coating of another material (e.g., titanium nitride, titanium carbonitride) to enhance wear resistance and tool life.

Tool Selection

Choosing the right cutting tool for a specific operation depends on several factors:

- **Workpiece Material:** Different tool materials are better suited for different workpiece materials.
- **Operation Type:** Turning, facing, boring, etc., each require specific tool geometries.
- **Cutting Conditions:** Cutting speed, feed rate, and depth of cut influence tool selection.
- **Desired Surface Finish:** The required surface finish dictates the tool nose radius and cutting edge angle.

Tool Maintenance

Proper tool maintenance is crucial for ensuring tool life and machining quality. This includes:

- **Sharpening:** Regularly sharpening cutting tools to maintain their cutting edge.
- **Chip Removal:** Removing chips from the tool and workpiece to prevent overheating and tool damage.
- **Lubrication and Cooling:** Using appropriate cutting fluids to reduce friction, heat, and wear.
- **Storage:** Storing tools properly to prevent damage and corrosion.

2.2 Workholding and Chucking

Workholding refers to securing the workpiece to the lathe so that it can be rotated and machined safely and accurately. The choice of workholding method depends on the shape, size, and material of the workpiece, as well as the specific machining operation.

Common Workholding Methods

- **Chuck:** A versatile device with jaws that grip the workpiece securely. Chucks come in various types, including three-jaw chucks for round workpieces and four-jaw chucks for irregular shapes.
- **Collet:** A type of chuck with a tapered bore that grips the workpiece tightly. Collets are known for their accuracy and concentricity.
- **Centers:** Workpieces with center holes can be supported between a live center (mounted on the spindle) and a dead center (mounted on the tailstock).
- **Faceplate:** A large, flat plate with slots or holes for attaching the workpiece. Faceplates are used for machining irregularly shaped workpieces or those that cannot be held in a chuck.
- **Mandrel:** A cylindrical bar with a tapered section that is inserted into the workpiece bore to support it during machining.

Chucking Techniques

Proper chucking is essential for ensuring workpiece stability and accuracy during machining. Here are some tips for chucking workpieces:

- **Centering:** Ensure the workpiece is centered in the chuck to avoid runout and vibration.
- **Tightening:** Tighten the chuck jaws evenly to avoid distortion of the workpiece.
- **Clearance:** Ensure sufficient clearance between the chuck jaws and the cutting tool to prevent interference.
- **Support:** For long workpieces, use the tailstock for additional support.

2.3 Measuring and Marking Techniques

Accurate measurement and marking are crucial for achieving precise machining results on the lathe. Several tools and techniques are used to measure and mark workpieces:

Measuring Tools

- **Calipers:** Used for measuring external and internal dimensions, depths, and thicknesses.
- **Micrometers:** Used for precise measurement of external dimensions.
- **Dial Indicators:** Used for measuring runout, alignment, and other dimensional variations.
- **Rulers and Scales:** Used for general measurements and layout.

Marking Techniques

- **Scriber:** A sharp-pointed tool used for scratching lines on the workpiece surface.
- **Layout Dye:** A colored liquid applied to the workpiece surface to enhance the visibility of scribed lines.
- **Center Punch:** A tool used to create a small indentation for starting drills and other tools.
- **Hermaphrodite Caliper:** A tool with one leg bent at 90 degrees, used for scribing lines parallel to an edge.

Measurement Tips

- **Zero Setting:** Always zero the measuring tool before taking a measurement.
- **Consistent Pressure:** Apply consistent pressure when taking measurements to ensure accuracy.
- **Multiple Measurements:** Take multiple measurements and average them to improve accuracy.
- **Reference Points:** Use reference points (e.g., chuck face, center line) to ensure consistent measurements.

2.4 Cutting Speeds and Feeds

Cutting speed and feed rate are critical parameters that significantly affect machining efficiency, tool life, and surface finish.

Cutting Speed

Cutting speed refers to the speed at which the cutting tool moves relative to the workpiece surface. It is expressed in surface feet per minute (SFM) or meters per minute (m/min). The optimal cutting speed depends on several factors:

- **Workpiece Material:** Harder materials require lower cutting speeds.
- **Tool Material:** Different tool materials have different recommended cutting speeds.
- **Tool Geometry:** Rake angle, clearance angle, and nose radius affect cutting speed.
- **Machine Power:** The lathe's power limits the maximum achievable cutting speed.

Feed Rate

Feed rate refers to the distance the cutting tool advances along the workpiece for each revolution of the spindle. It is expressed in inches per revolution (IPR) or millimeters per revolution (mm/r). The feed rate affects the chip thickness and surface finish. A higher feed rate removes more material per revolution but can result in a rougher surface finish.

Calculating Cutting Speed and Feed

Various formulas and charts are available for calculating the optimal cutting speed and feed rate for different materials and tools. It's essential to refer to these resources or consult with experienced machinists to determine the appropriate values for your specific setup.

2.5 Lubrication and Coolants

Lubrication and cooling play a crucial role in lathe operations by reducing friction, heat, and wear, improving tool life, and enhancing surface finish.

Types of Cutting Fluids

- **Cutting Oils:** Mineral or synthetic oils used for lubrication and cooling. They are suitable for heavy-duty machining operations.
- **Soluble Oils:** Emulsions of oil and water that offer both lubrication and cooling properties.
- **Semi-Synthetic and Synthetic Coolants:** Water-based fluids with additives to improve lubricity, cooling, and rust prevention.
- **Air Coolants:** Compressed air used for cooling in certain applications.

Applying Cutting Fluids

Cutting fluids can be applied to the cutting zone in various ways:

- **Flooding:** A continuous stream of fluid directed at the cutting zone.
- **Misting:** A fine mist of fluid sprayed onto the cutting zone.
- **Minimum Quantity Lubrication (MQL):** A small amount of fluid mixed with compressed air and applied directly to the cutting edge.

Choosing the Right Cutting Fluid

The choice of cutting fluid depends on several factors:

- **Workpiece Material:** Different materials require different types of cutting fluids.
- **Operation Type:** Turning, drilling, threading, etc., may require specific cutting fluids.
- **Cutting Speed and Feed:** Higher speeds and feeds generate more heat and require better cooling.
- **Environmental Concerns:** Consider the environmental impact of different cutting fluids.

2.6 Troubleshooting Common Problems

During lathe operations, you may encounter various issues that can affect machining quality or safety. Here are some common problems and their potential solutions:

Chatter: Vibrations during cutting caused by improper tool geometry, loose workholding, or excessive overhang. Solutions include adjusting tool geometry, tightening workholding, or reducing cutting parameters.

- **Poor Surface Finish:** Rough surface finish caused by dull tools, incorrect cutting speed or feed, or improper lubrication. Solutions include sharpening tools, adjusting cutting parameters, or using appropriate cutting fluids.
- **Taper:** Unintended taper in the workpiece caused by misalignment of the tailstock or tool

misalignment. Solutions include realigning the tailstock or correcting tool height.
- **Overheating:** Excessive heat generation during cutting caused by high cutting speed, insufficient lubrication, or dull tools. Solutions include reducing cutting speed, increasing lubrication, or sharpening tools.
- **Tool Breakage:** Tool breakage caused by excessive cutting forces, incorrect tool geometry, or improper tool material. Solutions include reducing cutting forces, correcting tool geometry, or using a more suitable tool material.

Regularly inspecting the machine, tools, and workpiece can help identify potential problems early on and prevent them from escalating into more significant issues.

2.7 Maintenance and Cleaning

Proper maintenance and cleaning of your metal lathe are essential for ensuring its longevity, accuracy, and safety. Regular maintenance tasks help prevent breakdowns, reduce wear and tear, and maintain optimal performance.

Daily Maintenance

- **Cleaning:** Remove chips and debris from the lathe bed, carriage, cross slide, and other components after each use.
- **Lubrication:** Lubricate the ways, lead screw, cross slide screw, and other moving parts according to the manufacturer's instructions.

- **Tool Inspection:** Inspect cutting tools for wear, damage, or dullness and replace or sharpen as needed.
- **Chuck Inspection:** Check the chuck jaws for proper alignment and tightness.
- **Tailstock Alignment:** Verify the tailstock alignment to ensure accurate turning between centers.

Weekly Maintenance

- **Way Cleaning and Lubrication:** Thoroughly clean and lubricate the lathe ways to prevent wear and ensure smooth movement.
- **Headstock Lubrication:** Lubricate the headstock bearings and gears according to the manufacturer's recommendations.
- **Motor Inspection:** Check the motor for proper operation and lubrication.
- **Electrical Connections:** Inspect electrical connections for tightness and corrosion.

Monthly Maintenance

- **Leveling:** Check the lathe's levelness and adjust if necessary to maintain accuracy.
- **Gib Adjustment:** Adjust the gibs on the carriage and cross slide to ensure smooth movement and eliminate play.
- **Belt Inspection:** Inspect belts for wear, tension, and alignment and replace or adjust as needed.
- **Coolant System:** Clean the coolant system and replace the coolant if necessary.

Troubleshooting and Repairs

If you encounter any issues with the lathe, refer to the manufacturer's troubleshooting guide or consult with a qualified technician. Do not attempt to repair complex problems yourself unless you have the necessary expertise and experience.

By following a regular maintenance schedule and promptly addressing any issues, you can ensure that your metal lathe remains a reliable and productive asset in your workshop for years to come.

CHAPTER 3
TURNING OPERATIONS

Turning is the foundational operation performed on a metal lathe, and it involves reducing the diameter of a workpiece by removing material as it rotates against a cutting tool. Through precise control of the cutting tool's movement and the workpiece's rotation, a wide array of shapes and features can be crafted on cylindrical workpieces. This chapter delves into the essential turning techniques, tooling, and considerations for both beginners and experienced machinists.

3.1 Straight Turning

Straight turning is the simplest turning operation, involving reducing a workpiece's diameter to a uniform size along its length. The cutting tool is fed parallel to the workpiece axis, creating a straight cylindrical surface. This technique is fundamental for creating shafts, pins, and other cylindrical components.

Tooling for Straight Turning

- **Roughing Tool:** A robust tool with a large nose radius and chipbreaker for efficient material removal in the initial stages.
- **Finishing Tool:** A sharp tool with a smaller nose radius for achieving a smooth and precise final surface.

Steps in Straight Turning

1. **Workpiece Preparation:** Ensure the workpiece is securely mounted in the chuck or between centers.
2. **Tool Selection:** Choose the appropriate roughing and finishing tools based on the workpiece material and desired finish.
3. **Tool Setting:** Set the cutting tool's height to the centerline of the workpiece and adjust the tool holder for proper clearance and rake angles.
4. **Roughing:** Make multiple passes with the roughing tool, gradually reducing the diameter until close to the final dimension.
5. **Finishing:** Make a final pass with the finishing tool to achieve the desired diameter and surface finish.
6. **Measurement:** Measure the diameter with calipers or a micrometer to verify accuracy.

3.2 Taper Turning

Taper turning involves creating a conical surface on the workpiece, where the diameter gradually changes along its length. Tapers can be external (on the outside diameter) or internal (on the inside diameter). They are essential for creating components like axles, tool shanks, and Morse tapers.

Methods of Taper Turning

- **Compound Rest:** The compound rest is set at an angle to feed the tool along the desired taper.

- **Taper Attachment:** A specialized attachment that guides the tool along a predetermined taper angle.
- **Offset Tailstock:** The tailstock is offset to create a taper when turning between centers.

Steps in Taper Turning

1. **Taper Calculation:** Determine the taper angle and length using trigonometry or a taper calculator.
2. **Tool Selection:** Choose the appropriate turning tool for the taper angle and workpiece material.
3. **Method Selection:** Select the most suitable method for creating the taper based on the workpiece and lathe capabilities.
4. **Setup:** Set up the compound rest, taper attachment, or tailstock offset as needed.
5. **Turning:** Feed the tool along the workpiece at the calculated angle to create the taper.
6. **Measurement:** Measure the taper diameter at different points to verify accuracy.

3.3 Facing

Facing involves creating a flat surface on the end of a workpiece. This operation is essential for squaring the ends of workpieces, preparing surfaces for subsequent operations, and creating features like shoulders and recesses.

Facing Tools

- **Facing Tool:** A tool with a cutting edge perpendicular to the workpiece axis, designed for facing operations.

Steps in Facing

1. **Workpiece Setup:** Secure the workpiece in the chuck with the end to be faced extending outward.
2. **Tool Selection:** Choose a suitable facing tool based on the workpiece material and desired finish.
3. **Tool Setting:** Set the cutting tool's height to the centerline of the workpiece and adjust the tool holder for proper clearance.
4. **Facing:** Feed the tool across the end of the workpiece, taking shallow cuts until the desired surface is achieved.
5. **Measurement:** Measure the faced surface for flatness and perpendicularity to the workpiece axis.

3.4 Parting Off

Parting off, also known as cutting off, involves separating a completed workpiece from the bar stock. It requires a specialized parting tool and careful technique to avoid tool breakage and ensure a clean cut.

Parting Tools

- **Parting Tool:** A narrow tool with a sharp cutting edge designed for cutting off workpieces.

Steps in Parting Off

1. **Workpiece Preparation:** Ensure the workpiece is securely held and the parting location is marked.
2. **Tool Selection:** Choose a suitable parting tool based on the workpiece material and diameter.
3. **Tool Setting:** Set the parting tool's height to the centerline of the workpiece and adjust the tool holder for proper clearance.
4. **Cutting:** Feed the parting tool into the workpiece slowly and steadily, applying cutting fluid as needed.
5. **Breakaway:** As the cut nears completion, reduce the feed rate to prevent the tool from jamming or breaking.
6. **Deburring:** Remove any burrs or sharp edges from the parted workpiece.

3.5 Knurling

Knurling creates a diamond-shaped or straight-line pattern on the surface of a workpiece. This pattern enhances grip, adds decorative detail, or provides a reference surface for measurements.

Knurling Tools

- **Knurling Tool:** A tool with one or more hardened wheels that imprint the knurling pattern onto the workpiece.

Steps in Knurling

1. **Workpiece Preparation:** Ensure the workpiece is securely held and the knurling location is marked.
2. **Tool Selection:** Choose the appropriate knurling tool based on the desired pattern and pitch.
3. **Tool Setting:** Set the knurling tool's height to the centerline of the workpiece and adjust the pressure.
4. **Knurling:** Feed the tool into the rotating workpiece, applying pressure until the desired pattern is achieved.
5. **Inspection:** Inspect the knurled surface for uniformity and depth.

3.6 Thread Cutting

Thread cutting involves creating helical grooves on the workpiece to form screw threads. Threads can be external (on the outside diameter) or internal (on the inside diameter). They are essential for creating fasteners, threaded connections, and various mechanical components.

Threading Tools

- **Taps:** Used for cutting internal threads.
- **Dies:** Used for cutting external threads.

- **Thread Chasers:** Used for cleaning up or repairing damaged threads.

Steps in Thread Cutting

1. **Thread Calculation:** Determine the thread size, pitch, and type (e.g., metric, Unified, Acme).
2. **Tool Selection:** Choose the appropriate tap or die based on the thread specifications.
3. **Workpiece Preparation:** Ensure the workpiece is securely held and the threading location is marked.
4. **Taping/Threading:** For internal threads, use a tap to cut the threads. For external threads, use a die.
5. **Lubrication:** Apply cutting fluid to the tap or die to reduce friction and prevent thread damage.
6. **Thread Inspection:** Inspect the threads for proper fit and finish using a thread gauge or mating part.

3.7 Advanced Turning Techniques

In addition to the basic turning operations covered above, there are several advanced techniques that experienced machinists can utilize to create complex shapes and achieve specific results:

- **Eccentric Turning:** Turning a workpiece with its axis offset from the center of rotation, creating cam-like profiles or non-circular shapes.

- **Form Turning:** Using a specialized form tool to create a specific profile or shape on the workpiece.
- **Plunge Cutting:** Feeding the tool radially into the workpiece to create grooves or cut off sections.
- **Contour Turning:** Combining turning and facing operations to create complex contours on the workpiece.

By mastering these advanced techniques, machinists can expand their creative capabilities and tackle a wider range of challenging projects.

CHAPTER 4
DRILLING AND BORING OPERATIONS

Drilling and boring are essential lathe operations that involve creating and enlarging holes in workpieces. These operations are critical for various applications, from simple through-holes to complex internal features. This chapter delves into the key techniques, tooling, and considerations for successful drilling and boring on a metal lathe.

4.1 Center Drilling

Center drilling is a preliminary operation used to create a small indentation, known as a center hole, at the end of a workpiece. This center hole serves as a starting point for subsequent drilling operations and helps guide the drill bit to ensure accurate hole placement and prevent it from wandering.

Center Drills

Center drills are specialized tools with a short, tapered point and two cutting edges. They are designed to create shallow center holes with a specific angle (usually 60 degrees) that matches the angle of standard lathe centers.

Steps in Center Drilling

1. **Workpiece Setup:** Secure the workpiece in the chuck or between centers.

2. **Tool Selection:** Choose a center drill with a diameter appropriate for the workpiece size and intended hole diameter.
3. **Tool Setting:** Set the center drill's point to the desired location on the workpiece end and adjust the tailstock for proper alignment.
4. **Drilling:** Slowly advance the center drill into the workpiece, using cutting fluid if necessary. Drill to a depth slightly greater than the drill bit's point angle.
5. **Withdrawal:** Withdraw the center drill carefully to avoid chipping or damaging the workpiece.

4.2 Drilling

Drilling is the process of creating a hole in a workpiece using a rotating drill bit. Drilling on a lathe is typically done with the workpiece held stationary in the chuck and the drill bit mounted in the tailstock. This method provides stability and control for accurate drilling.

Drill Bits

Twist drills are the most common type of drill bit used on lathes. They have a helical flute that helps remove chips and a pointed tip with two cutting edges. Other types of drill bits include spade drills, core drills, and gun drills, each with specific applications.

Steps in Drilling

1. **Workpiece Preparation:** Center drill the workpiece to ensure accurate hole placement.

2. **Tool Selection:** Choose the appropriate drill bit based on the desired hole diameter and workpiece material.
3. **Tool Setting:** Mount the drill bit in the tailstock drill chuck and align it with the center hole.
4. **Drilling:** Slowly advance the drill bit into the workpiece, applying cutting fluid as needed. Use a pecking motion (withdrawing the drill bit periodically) to clear chips and prevent overheating.
5. **Breakthrough:** As the drill bit nears breakthrough, reduce the feed rate to prevent chipping or grabbing.
6. **Withdrawal:** Withdraw the drill bit slowly and carefully to avoid damage to the workpiece or drill bit.

4.3 Reaming

Reaming is a finishing operation used to enlarge a drilled hole to a precise size and improve its surface finish. Reaming removes a small amount of material and leaves a smooth, accurate hole.

Reamers

Reamers are cutting tools with multiple cutting edges that are slightly larger than the drilled hole. They come in various types, including hand reamers, machine reamers, and chucking reamers.

Steps in Reaming

1. **Workpiece Preparation:** Ensure the drilled hole is clean and free of burrs.

2. **Tool Selection:** Choose a reamer with the correct size and tolerance for the desired hole finish.
3. **Tool Setting:** Mount the reamer in the tailstock or tool post and align it with the drilled hole.
4. **Reaming:** Slowly feed the reamer through the hole, using cutting fluid if necessary. Avoid excessive pressure to prevent tool breakage or damage to the workpiece.
5. **Withdrawal:** Withdraw the reamer carefully to avoid marring the finished surface.

4.4 Boring

Boring is the process of enlarging an existing hole in a workpiece. It is similar to drilling, but instead of creating a new hole, boring expands an existing one to a larger diameter. Boring is used to create precise internal features like cylinders, recesses, and undercuts.

Boring Tools

Boring bars are the primary tools used for boring operations. They consist of a long, rigid bar with a single-point cutting tool mounted on one end. Boring bars come in various sizes and configurations to accommodate different hole diameters and depths.

Steps in Boring

1. **Workpiece Preparation:** Secure the workpiece in the chuck or fixture.

2. **Tool Selection:** Choose a boring bar with the appropriate diameter and length for the hole to be bored.
3. **Tool Setting:** Mount the boring bar in the tool post and adjust the cutting tool's height to the centerline of the workpiece.
4. **Boring:** Slowly feed the boring bar into the existing hole, taking light cuts and using cutting fluid as needed.
5. **Measurement:** Measure the bored hole diameter to verify accuracy.

4.5 Counterboring and Countersinking

Counterboring and countersinking are operations used to create a stepped or tapered hole in a workpiece. Counterboring creates a larger diameter hole at the entrance of a smaller hole to accommodate the head of a bolt or screw. Countersinking creates a conical-shaped recess at the entrance of a hole to accommodate the tapered head of a screw.

Counterbores and Countersinks

Counterbores and countersinks are specialized cutting tools designed for their respective operations. They typically have multiple cutting edges and a pilot to guide them into the existing hole.

Steps in Counterboring and Countersinking

1. **Workpiece Preparation:** Ensure the workpiece is securely held and the hole to be counterbored or countersunk is drilled.

2. **Tool Selection:** Choose the appropriate counterbore or countersink based on the desired hole diameter and depth.
3. **Tool Setting:** Mount the tool in the tailstock or tool post and align it with the drilled hole.
4. **Operation:** Slowly feed the tool into the hole, taking light cuts and using cutting fluid as needed.
5. **Measurement:** Measure the counterbored or countersunk hole to verify dimensions and depth.

4.6 Thread Tapping

Thread tapping is the process of cutting internal threads in a hole using a tap. Tapping is essential for creating threaded connections for bolts, screws, and other fasteners.

Taps

Taps are cutting tools with multiple flutes and cutting edges designed to create internal threads. They come in various types, including hand taps, machine taps, and spiral taps.

Steps in Thread Tapping

1. **Workpiece Preparation:** Drill the hole to the correct tap drill size, which is slightly smaller than the desired thread diameter.
2. **Tool Selection:** Choose the appropriate tap based on the thread specifications and workpiece material.

3. **Tool Setting:** Mount the tap in a tap wrench or tapping head and align it with the drilled hole.
4. **Tapping:** Turn the tap clockwise while applying light pressure to cut the threads. Use cutting fluid to reduce friction and prevent thread damage.
5. **Chip Removal:** Periodically reverse the tap to break and remove chips.
6. **Thread Inspection:** Inspect the threads for proper fit and finish using a thread gauge or mating part.

4.7 Honing

Honing is a precision finishing process used to improve the surface finish and accuracy of a hole. Honing stones, which are abrasive tools with a fine grit, are used to remove a small amount of material and create a smooth, cylindrical surface.

Honing Tools

Honing machines or portable honing tools are used for honing operations. These tools typically have a rotating mandrel with honing stones attached to it.

Steps in Honing

1. **Workpiece Preparation:** Ensure the hole to be honed is clean and free of burrs.
2. **Tool Selection:** Choose honing stones with the appropriate grit and size for the desired surface finish and hole diameter.

3. **Tool Setting:** Mount the honing tool in the tailstock or machine spindle and align it with the hole.
4. **Honing:** Slowly rotate and reciprocate the honing tool within the hole, applying light pressure and using honing oil as a lubricant.
5. **Measurement:** Measure the honed hole diameter and surface finish to verify accuracy.

By understanding the various drilling and boring operations, machinists can effectively create and modify holes in workpieces to meet specific requirements. Proper tool selection, setup, and technique are crucial for achieving accurate and high-quality results. With practice and attention to detail, machinists can master these essential lathe operations and expand their capabilities in metalworking.

CHAPTER 5

THREADING OPERATIONS

Threading is the process of creating helical grooves on a workpiece to form screw threads. These threads enable the assembly and disassembly of various components and play a crucial role in countless mechanical applications. This chapter delves into the intricacies of threading operations, covering external and internal threading techniques, thread repair, tooling selection, measurement, and troubleshooting.

5.1 External Threading

External threading involves cutting threads on the outside diameter of a cylindrical workpiece. This process is commonly used to create bolts, screws, studs, and other threaded fasteners. Several methods are employed for external threading, each with its own advantages and limitations.

Methods of External Threading

- **Single-Point Threading:** This traditional method utilizes a single-point threading tool, which is ground to the specific thread form. The tool is fed along the workpiece at the correct pitch to create the thread. This method offers flexibility and is suitable for various thread types, but requires skill and precision.
- **Die Threading:** Dies are specialized tools with multiple cutting teeth that form the thread profile. The workpiece is rotated while the die is advanced along the outside diameter,

creating the thread. Die threading is faster and more efficient than single-point threading, but is limited to specific thread sizes and types.
- **Thread Milling:** This advanced technique uses a rotating milling cutter to create the thread profile. Thread milling offers high precision and can produce complex thread forms, but requires specialized equipment and tooling.

Tooling for External Threading

- **Single-Point Threading Tools:** These tools come in various shapes and sizes to accommodate different thread forms and pitches.
- **Dies:** Available in round, hex, and adjustable designs, dies are suitable for specific thread sizes and types.
- **Thread Milling Cutters:** These cutters are designed for specific thread forms and pitches and can be used on milling machines or lathes with milling attachments.

Steps in External Threading

1. **Thread Calculation:** Determine the thread size, pitch, and type (e.g., metric, Unified, Acme).
2. **Workpiece Preparation:** Ensure the workpiece is securely held and the threading starting point is marked.
3. **Tool Selection:** Choose the appropriate threading tool based on the thread specifications and workpiece material.

4. **Tool Setting:** Set the threading tool's height to the centerline of the workpiece and align it with the thread starting point.
5. **Threading:** Feed the tool along the workpiece at the correct pitch and depth of cut, applying cutting fluid as needed. Use a thread chasing dial or other threading aid to ensure accurate thread lead.
6. **Thread Inspection:** Inspect the threads for proper fit and finish using a thread gauge or mating part.

5.2 Internal Threading

Internal threading involves cutting threads on the inside diameter of a hole or bore. This process is commonly used to create nuts, threaded holes, and other internal threaded components. Similar to external threading, internal threading can be achieved through various methods and tools.

Methods of Internal Threading

- **Tapping:** The most common method, tapping utilizes a tap, a cutting tool with multiple flutes and cutting edges, to create internal threads. The tap is rotated and fed into a pre-drilled hole to form the thread.
- **Single-Point Threading:** This method involves using a single-point threading tool to cut internal threads. It requires skill and precision but offers flexibility for different thread forms and sizes.
- **Internal Thread Milling:** Similar to external thread milling, this technique uses a rotating milling cutter to create internal threads. It

offers high precision and can produce complex thread forms, but requires specialized equipment and tooling.

Tooling for Internal Threading

- **Taps:** Available in various types, including hand taps, machine taps, and spiral taps, for different thread sizes and materials.
- **Single-Point Threading Tools:** These tools are similar to those used for external threading, but with a longer shank to reach into the hole.
- **Internal Thread Milling Cutters:** Designed for specific thread forms and pitches, these cutters are used on milling machines or lathes with milling attachments.

Steps in Internal Threading

1. **Hole Preparation:** Drill a hole with the correct tap drill size, which is slightly smaller than the desired thread diameter.
2. **Tool Selection:** Choose the appropriate tap or threading tool based on the thread specifications and workpiece material.
3. **Tool Setting:** Mount the tap in a tap wrench or tapping head, or set up the single-point threading tool in the tool post.
4. **Threading:** Rotate the tap or tool clockwise while applying gentle pressure to cut the threads. Use cutting fluid to reduce friction and prevent thread damage.
5. **Chip Removal:** Periodically reverse the tap or tool to break and remove chips.

6. **Thread Inspection:** Inspect the threads for proper fit and finish using a thread plug gauge or mating part.

5.3 Thread Repair

Thread repair is the process of restoring damaged or worn threads to their original condition. This can involve cleaning up minor imperfections, recutting damaged threads, or inserting threaded inserts to reinforce or replace worn-out threads.

Thread Repair Methods

- **Thread Chasers:** Used to clean up burrs, minor imperfections, and lightly damaged threads.
- **Taps and Dies:** Used to recut damaged threads to their original size and pitch.
- **Thread Inserts:** Threaded inserts, such as Helicoil or Keensert, are inserted into a drilled hole to provide a new, strong thread.

Steps in Thread Repair

1. **Assessment:** Assess the extent of the thread damage and determine the most suitable repair method.
2. **Preparation:** Clean the damaged threads of any debris or rust.
3. **Repair:** Use the appropriate tool or insert to restore the threads to their original condition.
4. **Inspection:** Inspect the repaired threads for proper fit and finish using a thread gauge or mating part.

5.4 Choosing the Right Taps and Dies

Selecting the correct taps and dies is crucial for successful threading operations. Several factors should be considered when choosing these tools:

- **Thread Type:** Ensure the tap or die matches the desired thread type (e.g., metric, Unified, Acme).
- **Thread Size and Pitch:** Choose the correct size and pitch of the tap or die to match the desired thread specifications.
- **Material:** Different tap and die materials are suitable for different workpiece materials. High-speed steel (HSS) is a common choice for general-purpose threading, while carbide is preferred for harder materials.
- **Coating:** Coated taps and dies offer improved wear resistance and lubricity, extending their life and improving thread quality.
- **Quality:** Invest in high-quality taps and dies from reputable manufacturers to ensure accuracy, durability, and consistent results.

5.5 Thread Measurement and Inspection

Accurate measurement and inspection of threads are essential for ensuring proper fit and function of threaded components. Several tools and techniques are used to measure and inspect threads:

- **Thread Gauges:** Go and no-go thread gauges are used to verify the accuracy of external and internal threads.
- **Thread Micrometers:** Used to measure the pitch diameter of threads with high precision.

- **Thread Ring Gauges:** Used to measure the major diameter of external threads.
- **Thread Plug Gauges:** Used to measure the minor diameter of internal threads.
- **Optical Comparators:** Used to visually inspect thread profiles for accuracy and consistency.
- **Thread Roll Snaps:** Used to check the fit of external threads by attempting to screw them onto a threaded gauge.

5.6 Threading Accessories

Various accessories can aid in threading operations, improving efficiency, accuracy, and safety. Some common threading accessories include:

- **Tap Wrenches:** Used to turn taps during internal threading.
- **Die Stocks:** Used to hold and turn dies during external threading.
- **Tapping Heads:** Adjustable tapping heads can be mounted on the tailstock or tool post for precise tapping operations.
- **Thread Chasing Dials:** These dials help maintain accurate thread lead during single-point threading.
- **Thread Cutting Oil:** Special lubricants designed to reduce friction and heat during threading, improving thread quality and tool life.

5.7 Troubleshooting Threading Issues

Threading operations can sometimes present challenges, resulting in poor thread quality, tool breakage, or other problems. Here are some common threading issues and their potential solutions:

- **Torn Threads:** This can be caused by dull tools, incorrect tap drill size, insufficient lubrication, or excessive cutting pressure. Solutions include using sharp tools, selecting the correct tap drill size, applying adequate lubrication, and reducing cutting pressure.
- **Poor Thread Finish:** Rough or uneven threads can be caused by dull tools, incorrect cutting speed or feed, or improper tool alignment. Solutions include sharpening tools, adjusting cutting parameters, and ensuring proper tool alignment.
- **Broken Taps:** Taps can break due to excessive torque, misalignment, or hitting the bottom of a blind hole. Solutions include using a tapping head to prevent overtightening, ensuring proper alignment, and using a bottoming tap for blind holes.
- **Stripped Threads:** Stripped threads can occur when excessive force is applied to a threaded fastener. Solutions include using a thread insert to repair the damaged threads or replacing the component.

By understanding the principles of threading, choosing the right tools and accessories, and applying proper techniques, machinists can produce high-quality threads that meet the demands of various applications.

CHAPTER 6

FACING AND PARTING OPERATIONS

Facing and parting are fundamental lathe operations that involve creating flat surfaces on the ends of workpieces and separating them from the stock material. These operations are essential for preparing workpieces for subsequent machining, creating specific features, and producing finished components. This chapter delves into the techniques, tools, and safety considerations associated with facing and parting operations on a metal lathe.

6.1 Facing Techniques

Facing involves creating a flat, smooth surface on the end of a workpiece perpendicular to its axis. This operation is crucial for squaring off stock material, establishing reference surfaces for subsequent machining operations, and producing features like shoulders, steps, and recesses.

Facing Tool Geometry

The cutting edge of a facing tool is typically perpendicular to the workpiece axis, enabling it to remove material across the end face in a single pass. The tool's nose radius, side rake angle, and back rake angle are important factors influencing the cutting forces, chip formation, and surface finish.

Facing Methods

- **Conventional Facing:** This method involves feeding the facing tool across the end of the workpiece in a single, continuous motion. It is suitable for creating flat surfaces on workpieces with relatively small diameters.
- **Plunge Facing:** For larger workpieces, plunge facing is preferred. The tool is first plunged into the center of the workpiece, then fed radially outward to the edge, creating a flat surface in a series of concentric cuts. This method minimizes tool deflection and ensures a more even surface finish.

Steps in Facing

1. **Workpiece Setup:** Secure the workpiece in the chuck or between centers with the end to be faced extending outward.
2. **Tool Selection:** Choose a facing tool with the appropriate size and geometry for the workpiece material and desired finish.
3. **Tool Setting:** Set the cutting tool's height to the centerline of the workpiece and adjust the tool holder for proper clearance and rake angles.
4. **Facing:** Feed the tool across the end of the workpiece, taking shallow cuts until the desired surface is achieved. For plunge facing, start at the center and gradually move outward.
5. **Measurement:** Measure the faced surface for flatness and perpendicularity to the workpiece axis using a dial indicator or square.

6.2 Parting Off

Parting off, also known as cutting off, is the process of separating a completed workpiece from the bar stock. This operation requires a specialized parting tool and careful technique to achieve a clean, precise cut without damaging the workpiece or tool.

Parting Tool Geometry

A parting tool has a narrow blade with a sharp cutting edge that is fed into the workpiece perpendicular to its axis. The tool's width, top rake angle, and side clearance angles are critical factors influencing its performance and ability to create a clean cut.

Steps in Parting Off

1. **Workpiece Preparation:** Ensure the workpiece is securely held and the parting location is marked.
2. **Tool Selection:** Choose a parting tool with the appropriate width and cutting edge geometry for the workpiece material and diameter.
3. **Tool Setting:** Set the parting tool's height to the centerline of the workpiece and adjust the tool holder for proper clearance and rake angles.
4. **Cutting:** Feed the parting tool into the workpiece slowly and steadily, applying cutting fluid as needed. Maintain consistent feed pressure to avoid tool chatter or breakage.
5. **Breakaway:** As the cut nears completion, reduce the feed rate and support the workpiece with the tailstock center or a follower rest to

prevent the tool from jamming or pulling the workpiece out of the chuck.
6. **Deburring:** Remove any burrs or sharp edges from the parted workpiece using a deburring tool or file.

6.3 Grooving

Grooving involves creating a narrow channel or recess on the workpiece surface. Grooves can be external (on the outside diameter) or internal (on the inside diameter) and are used for various purposes, such as retaining rings, O-rings, or creating decorative features.

Grooving Tools

- **Grooving Tool:** A narrow tool with a single cutting edge designed for creating grooves. The tool's width determines the width of the groove.
- **Form Tool:** For complex groove profiles, a form tool with the desired shape can be used.

Steps in Grooving

1. **Workpiece Preparation:** Secure the workpiece in the chuck or between centers and mark the groove location.
2. **Tool Selection:** Choose the appropriate grooving tool or form tool based on the groove width and profile.
3. **Tool Setting:** Set the tool's height to the desired groove depth and adjust the tool holder for proper clearance and rake angles.

4. **Grooving:** Feed the tool into the workpiece to the desired depth, using a steady feed rate and applying cutting fluid as needed. For external grooves, feed the tool radially. For internal grooves, feed the tool axially.
5. **Measurement:** Measure the groove width and depth to verify accuracy.

6.4 Form Cutting

Form cutting involves creating a specific profile or shape on the workpiece using a form tool. Form tools have a cutting edge that matches the desired profile, allowing for the production of complex shapes like gears, threads, or custom profiles.

Form Tools

- **Circular Form Tools:** Used for creating rounded profiles.
- **Dovetail Form Tools:** Used for cutting dovetail shapes.
- **Thread Cutting Form Tools:** Used for cutting threads with specific forms and pitches.
- **Custom Form Tools:** Can be made to create unique or specialized profiles.

Steps in Form Cutting

1. **Workpiece Preparation:** Secure the workpiece in the chuck or between centers.
2. **Tool Selection:** Choose a form tool with the desired profile and cutting edge geometry.
3. **Tool Setting:** Set the form tool's height to the centerline of the workpiece and adjust the tool holder for proper clearance and rake angles.

4. **Form Cutting:** Feed the form tool into the rotating workpiece, following the desired profile.
5. **Measurement:** Measure the formed profile to verify accuracy and match the desired specifications.

6.5 Cutting Off Bar Stock

Cutting off bar stock is similar to parting off a finished workpiece, but it typically involves cutting through a larger diameter piece of material. The same principles and techniques apply, but extra care is needed to ensure safe and efficient cutting.

Steps in Cutting Off Bar Stock

1. **Workpiece Preparation:** Secure the bar stock in the chuck with sufficient overhang for cutting.
2. **Tool Selection:** Choose a parting tool with the appropriate width and cutting edge geometry for the bar stock material and diameter.
3. **Tool Setting:** Set the parting tool's height to the centerline of the bar stock and adjust the tool holder for proper clearance and rake angles.
4. **Cutting:** Feed the parting tool into the bar stock slowly and steadily, applying cutting fluid as needed. Use a slower feed rate for larger diameters and harder materials.
5. **Breakaway:** As the cut nears completion, reduce the feed rate and support the bar stock with a steady rest or follower rest to prevent

the tool from jamming or pulling the bar stock out of the chuck.
6. **Deburring:** Remove any burrs or sharp edges from the cut end of the bar stock.

6.6 Facing and Parting Tools

The tools used for facing and parting operations are crucial for achieving accurate and efficient results. The choice of tool depends on the workpiece material, size, and desired finish.

Facing Tools

- **Left-Hand and Right-Hand Facing Tools:** These tools have cutting edges on the left or right side, allowing for facing towards or away from the headstock.
- **Centerline Facing Tools:** These tools have a cutting edge on both sides, enabling facing in either direction.
- **Shell End Mills:** These tools can be used for facing and profiling operations on larger workpieces.

Parting Tools

- **Narrow Parting Tools:** Used for cutting off smaller diameter workpieces and creating narrow grooves.
- **Wide Parting Tools:** Used for cutting off larger diameter workpieces and creating wider grooves.
- **Thinbit Parting Tools:** These inserts are designed for minimizing material waste and producing a clean cut.

6.7 Safety Considerations

Facing and parting operations can be hazardous if proper safety precautions are not followed. Here are some essential safety considerations:

- **Secure Workholding:** Ensure the workpiece is securely held in the chuck or between centers to prevent it from coming loose during cutting.
- **Tool Clearance:** Check for adequate clearance between the tool holder and the workpiece to avoid collisions.
- **Chip Removal:** Regularly remove chips from the cutting area to prevent them from interfering with the cutting process or causing overheating.
- **Breakaway:** When parting off, use a slower feed rate as the cut nears completion and support the workpiece to prevent tool breakage or workpiece damage.
- **Personal Protective Equipment (PPE):** Always wear safety glasses and other appropriate PPE to protect yourself from flying chips and debris.

By understanding the techniques, tools, and safety considerations outlined in this chapter, you can confidently and safely perform facing and parting operations on your metal lathe. With practice and attention to detail, you can achieve precise results and expand your machining capabilities.

CHAPTER 7

PROJECTS FOR BEGINNERS

Embarking on practical projects is an exciting and effective way for beginners to solidify their understanding of metal lathe operations and gain hands-on experience. This chapter presents a selection of engaging projects designed to introduce fundamental techniques while creating functional and aesthetically pleasing items.

7.1 Making a Simple Bushing

A bushing is a cylindrical sleeve that reduces friction between moving parts. It's a simple yet versatile project ideal for mastering basic turning and facing operations.

Materials and Tools:

- Metal rod (brass, aluminum, or steel) of appropriate diameter
- Lathe chuck
- Center drill
- Twist drill (size depending on the desired bushing inner diameter)
- Boring bar (optional)
- Reamer (optional)
- Cutting tools (roughing and finishing tools)
- Calipers or micrometer
- Deburring tool

Steps:

1. **Workpiece Preparation:** Cut the metal rod to the desired bushing length. Face both ends of the rod to ensure they are square and smooth.
2. **Center Drilling:** Center drill both ends of the rod to establish starting points for drilling.
3. **Drilling:** Drill a hole through the center of the rod using a twist drill of the appropriate size. If a more precise hole is required, use a boring bar to enlarge the hole and then finish it with a reamer.
4. **Turning:** Mount the rod in the chuck with the drilled hole facing outward. Turn the outer diameter to the desired bushing size, using roughing and finishing tools to achieve a smooth finish.
5. **Parting Off:** Part the finished bushing from the rod using a parting tool.
6. **Deburring:** Remove any burrs or sharp edges from the bushing using a deburring tool or file.

7.2 Turning a Pen Blank

Turning a pen blank is a rewarding project that allows you to create a personalized writing instrument. It involves shaping the pen blank to fit a pen kit, which includes all the necessary hardware for assembling the pen.

Materials and Tools:

- Pen blank (wood, acrylic, or other suitable material)
- Pen kit

- Lathe chuck or mandrel
- Center drill (if using a mandrel)
- Twist drill (size depending on the pen kit)
- Turning tools (roughing, finishing, and profiling tools)
- Calipers or micrometer
- Sandpaper (various grits)
- Pen finish (optional)

Steps:

1. **Workpiece Preparation:** Cut the pen blank to the appropriate length for the pen kit. Square the ends and drill a center hole if using a mandrel.
2. **Mounting:** Mount the pen blank in the chuck or onto the mandrel.
3. **Turning:** Turn the pen blank to the desired shape, using different tools to create various profiles and details.
4. **Sanding:** Sand the turned pen blank through progressively finer grits to achieve a smooth finish.
5. **Finishing:** Apply a pen finish (e.g., lacquer, wax) to protect and enhance the appearance of the pen blank.
6. **Assembly:** Assemble the pen according to the pen kit instructions.

7.3 Creating a Candlestick Holder

A candlestick holder is a simple yet elegant project that can be customized with various designs and finishes. It's a great way to practice turning, facing, and drilling operations.

Materials and Tools:

- Metal rod (brass, aluminum, or steel) of appropriate diameter
- Lathe chuck
- Center drill
- Twist drill (size depending on the candle diameter)
- Turning tools (roughing and finishing tools)
- Parting tool
- Calipers or micrometer
- Deburring tool
- Polishing compound (optional)

Steps:

1. **Workpiece Preparation:** Cut the metal rod to the desired candlestick holder length. Face both ends of the rod to ensure they are square and smooth.
2. **Base:** Turn one end of the rod to create a wider base for stability.
3. **Center Drilling:** Center drill the opposite end of the rod to establish a starting point for drilling.
4. **Drilling:** Drill a hole in the center of the rod to accommodate the candle.
5. **Turning:** Turn the outer diameter of the rod to the desired shape, creating any decorative details or profiles.
6. **Parting Off:** Part the finished candlestick holder from the rod.

7. **Deburring and Polishing:** Remove any burrs or sharp edges and polish the holder to a desired finish.

7.4 Machining a Bottle Opener

A bottle opener is a practical and fun project that involves turning, facing, and profiling operations. It's a great way to practice using different tools and techniques.

Materials and Tools:

- Metal bar stock (steel or stainless steel)
- Lathe chuck
- Center drill
- Twist drill
- Turning tools (roughing, finishing, and profiling tools)
- Hacksaw or bandsaw
- File
- Sandpaper
- Polishing compound

Steps:

1. **Workpiece Preparation:** Cut a piece of bar stock to the desired bottle opener length using a hacksaw or bandsaw.
2. **Facing and Center Drilling:** Face both ends of the bar stock and center drill one end.
3. **Turning:** Mount the bar stock in the chuck and turn the outer diameter to the desired

shape. Use profiling tools to create the bottle opener notch and any decorative features.
4. **Parting Off:** Part the finished bottle opener from the bar stock.
5. **Finishing:** File and sand the bottle opener to remove any sharp edges and achieve a smooth finish. Polish the opener with a polishing compound for a shiny look.

7.5 Building a Small Engine Piston

Machining a small engine piston is a more advanced project that requires precision and understanding of engine components. It's a great way to apply various lathe operations and gain experience with complex geometries.

Materials and Tools:

- Aluminum bar stock (specific alloy for engine pistons)
- Lathe chuck
- Center drill
- Twist drills (various sizes)
- Boring bar
- Turning tools (roughing, finishing, and profiling tools)
- Calipers or micrometer
- Piston ring grooves tool
- Micrometer or dial bore gauge
- Deburring tool
- Polishing compound

Steps:

1. **Workpiece Preparation:** Cut the aluminum bar stock to the desired piston length. Face both ends and center drill one end.
2. **Turning:** Mount the bar stock in the chuck and turn the outer diameter to the desired piston size, leaving extra material for finishing.
3. **Boring:** Bore the inside of the piston to create the wrist pin bore.
4. **Piston Ring Grooves:** Cut the piston ring grooves using a specialized tool or a grooving tool.
5. **Piston Crown:** Shape the piston crown using turning and profiling tools.
6. **Finishing:** Finish the piston to the final dimensions, ensuring proper clearances for the piston rings and wrist pin. Deburr and polish the piston for a smooth finish.

7.6 Project Ideas for Inspiration

The projects listed above are just a starting point. Many other exciting and challenging projects can be tackled with a metal lathe, depending on your interests and skill level:

- **Yo-yos:** Create custom yo-yos with various designs and materials.
- **Chess Pieces:** Turn and carve unique chess pieces from wood or metal.
- **Knobs and Handles:** Machine custom knobs and handles for tools, cabinets, or furniture.
- **Gear Blanks:** Turn and cut gear blanks for mechanical projects.

- **Jewelry:** Create rings, pendants, or other jewelry pieces.

The possibilities are endless, and with creativity and practice, you can use your metal lathe to bring your ideas to life.

7.7 Troubleshooting and Refinement

As you work on these projects, you may encounter challenges or imperfections in your work. Here are some common troubleshooting tips:

- **Chatter:** If you experience chatter during turning, check for loose toolholders, worn bearings, or incorrect tool geometry.
- **Poor Surface Finish:** A rough surface finish can be caused by dull tools, incorrect cutting speed or feed, or improper lubrication. Sharpen your tools, adjust cutting parameters, and use appropriate cutting fluids.
- **Dimensional Inaccuracies:** If your dimensions are not accurate, double-check your measurements and tool settings. Make sure the workpiece is securely held and the tailstock is properly aligned.

Remember, practice makes perfect. Don't be discouraged by mistakes; learn from them and continue to refine your techniques. As you gain experience, you'll be able to create more complex and impressive projects with your metal lathe.

CHAPTER 8

INTERMEDIATE LATHE TECHNIQUES

As you gain proficiency with basic lathe operations, exploring intermediate techniques opens a world of possibilities for creating more intricate and functional parts. This chapter delves into eccentric turning, taper attachments, milling attachments, indexing and dividing heads, compound rest operations, advanced threading, and the realm of creative lathe projects.

8.1 Eccentric Turning

Eccentric turning involves machining a workpiece where the axis of rotation is offset from the geometric center. This technique produces non-circular shapes, cam profiles, and other eccentric features that have applications in various mechanisms and machines.

Methods of Eccentric Turning

- **Four-Jaw Chuck:** By positioning the workpiece off-center in a four-jaw chuck, eccentric turning can be achieved. This method offers flexibility but requires careful setup and adjustment.
- **Eccentric Mandrel:** An eccentric mandrel is a specialized tool with an offset axis that holds the workpiece. This simplifies the setup process and ensures consistent eccentricity.

- **Offset Tailstock:** For workpieces turned between centers, offsetting the tailstock creates eccentricity.

Steps in Eccentric Turning

1. **Setup:** Secure the workpiece using the chosen method, ensuring the desired offset is achieved.
2. **Tool Selection:** Select the appropriate turning tool for the desired profile and material.
3. **Turning:** With the lathe running, carefully feed the cutting tool into the workpiece, following the desired profile. Maintain consistent feed and cutting depth for a smooth finish.
4. **Measurement:** Verify the eccentricity and profile accuracy using dial indicators and calipers.

8.2 Taper Attachment

A taper attachment is a versatile accessory that simplifies the creation of tapered surfaces on a lathe. It consists of a guide bar that can be set at various angles, allowing the tool to be fed along the desired taper automatically.

Types of Taper Attachments

- **Plain Taper Attachment:** This basic attachment provides a fixed taper angle and is suitable for most common taper turning operations.
- **Telescopic Taper Attachment:** This attachment allows for adjustable taper angles

within a certain range, offering greater flexibility.

Using a Taper Attachment

1. **Setup:** Attach the taper attachment to the lathe bed and set the guide bar to the desired taper angle.
2. **Tool Setting:** Align the cutting tool with the workpiece and set the depth of cut.
3. **Turning:** Engage the taper attachment and feed the tool along the workpiece. The taper attachment will automatically guide the tool to create the desired taper.
4. **Measurement:** Verify the taper angle and length using a taper gauge or other measuring tools.

8.3 Milling Attachments

A milling attachment is a valuable accessory that expands the capabilities of a lathe by enabling milling operations. It typically consists of a milling head, which holds the milling cutter, and a mounting bracket that attaches to the lathe's cross slide or compound rest.

Types of Milling Operations on a Lathe

- **Face Milling:** Creating flat surfaces on the ends of workpieces.
- **Slot Milling:** Cutting slots or keyways in workpieces.
- **Gear Cutting:** Creating gear teeth on cylindrical blanks.

- **Helical Milling:** Cutting helical grooves or threads on cylindrical workpieces.

Using a Milling Attachment

1. **Setup:** Attach the milling attachment to the lathe and mount the desired milling cutter.
2. **Workpiece Setup:** Secure the workpiece in the chuck or fixture.
3. **Tool Setting:** Set the milling cutter's position and adjust the spindle speed and feed rate.
4. **Milling:** Slowly feed the workpiece into the rotating milling cutter, using cutting fluid as needed.
5. **Measurement:** Verify the dimensions and accuracy of the milled features using appropriate measuring tools.

8.4 Indexing and Dividing Head

Indexing and dividing heads are essential accessories for precise angular positioning of workpieces on a lathe. They enable the division of a circle into equal parts, allowing for accurate drilling of bolt circles, cutting of gears, and other operations requiring precise angular spacing.

Types of Indexing and Dividing Heads

- **Simple Indexing Head:** This basic head allows for direct indexing of common divisions like 2, 3, 4, 6, 8, 12, and 24.
- **Universal Dividing Head:** This versatile head offers a wider range of indexing options and can be used for both simple and compound indexing.

Using an Indexing and Dividing Head

1. **Setup:** Attach the indexing or dividing head to the lathe's cross slide or milling attachment.
2. **Workpiece Setup:** Secure the workpiece in the chuck or fixture.
3. **Indexing:** Determine the desired number of divisions and use the indexing plate or dial to position the workpiece accurately.
4. **Operation:** Perform the desired operation (e.g., drilling, milling) at each indexed position.

8.5 Compound Rest Operations

The compound rest is a versatile component of a lathe that allows for angular movement of the cutting tool. This enables the machining of tapers, angles, chamfers, and other complex features.

Compound Rest Adjustments

- **Compound Angle:** The angle at which the compound rest is set relative to the lathe axis.
- **Tool Height:** The vertical position of the cutting tool relative to the workpiece centerline.
- **Feed Rate:** The rate at which the cutting tool is fed along the workpiece.

Compound Rest Operations

- **Taper Turning:** Creating tapered surfaces by setting the compound rest at the desired taper angle.

- **Angle Cutting:** Machining angles on workpieces by setting the compound rest to the required angle.
- **Chamfering:** Creating a beveled edge on the end of a workpiece using the compound rest.
- **Dovetail Cutting:** Machining dovetail joints for precise interlocking of components.

8.6 Advanced Threading

Beyond the basic threading operations covered in Chapter 5, advanced threading techniques allow for the creation of more complex and specialized thread forms.

Advanced Threading Techniques

- **Multi-Start Threads:** Threads with multiple helical grooves starting at different points around the circumference, offering faster travel and increased load-bearing capacity.
- **Acme Threads:** Trapezoidal-shaped threads known for their strength and wear resistance, commonly used in power transmission applications.
- **Buttress Threads:** Asymmetrical threads designed for high axial loads in one direction, used in applications like jacks and vices.

Tooling for Advanced Threading

- **Multi-Start Threading Tools:** Specialized tools for cutting multi-start threads.
- **Acme Threading Tools:** Designed for cutting Acme threads with the correct profile and pitch.

- **Buttress Threading Tools:** These tools create the unique asymmetrical profile of buttress threads.

8.7 Creative Lathe Projects

Once you have mastered the intermediate lathe techniques, you can explore a wide range of creative projects that push the boundaries of your skills and imagination. Here are some ideas to inspire you:

- **Artistic Turnings:** Create intricate and decorative wood or metal turnings with complex patterns and textures.
- **Custom Tooling:** Design and manufacture your own specialized tools and accessories for the lathe.
- **Mechanical Models:** Build working models of engines, clocks, or other mechanical devices.
- **Musical Instruments:** Craft unique instruments like flutes, ocarinas, or drums from various materials.
- **Sculptures:** Turn and carve sculptures from wood, metal, or other materials.

The possibilities are endless, and with creativity and experimentation, you can use your lathe to create truly unique and personalized works of art.

By delving into these intermediate lathe techniques, you'll significantly expand your machining capabilities and open new avenues for creativity and problem-solving.

CHAPTER 9

ADVANCED LATHE OPERATIONS

Venturing into advanced lathe operations unlocks a realm of precision, complexity, and creative possibilities. This chapter delves into the intricacies of CNC lathe programming, live tooling, multi-axis turning, gear and spline cutting, custom tooling, and the pursuit of high-precision machining.

9.1 CNC Lathe Programming

Computer Numerical Control (CNC) lathes revolutionize machining by automating toolpaths and operations through programmed instructions. This technology offers unparalleled precision, repeatability, and efficiency, making it indispensable for modern manufacturing.

G-Code Programming

G-code is the language used to communicate with CNC machines. It comprises a series of alphanumeric codes that instruct the machine on movements, spindle speeds, tool changes, and other parameters. CNC lathe programming involves writing G-code programs that define the precise sequence of operations required to machine a part.

Programming Concepts

- **Coordinate Systems:** CNC lathes typically use a Cartesian coordinate system with X and Z axes. The X-axis represents the diameter of the workpiece, while the Z-axis represents its length.
- **Toolpaths:** Toolpaths are the paths that the cutting tool follows to machine the workpiece. They can be linear, circular, or complex curves.
- **Cutting Parameters:** Cutting speed, feed rate, and depth of cut are critical parameters that must be carefully selected for each operation.
- **Tool Offsets:** Tool offsets compensate for differences in tool lengths and diameters, ensuring accurate machining.

CNC Lathe Programming Software

Various software packages are available for CNC lathe programming. These programs offer graphical interfaces, simulation tools, and post-processors that generate G-code specific to the machine controller. Popular CNC lathe programming software includes:

- **Mastercam:** A comprehensive CAM software with powerful lathe programming capabilities.
- **FeatureCAM:** A feature-based CAM software that simplifies programming complex parts.
- **GibbsCAM:** A user-friendly CAM software with a focus on ease of use and productivity.

Advantages of CNC Lathe Programming

- **Precision and Accuracy:** CNC lathes achieve high levels of precision and accuracy, ensuring consistent results across multiple parts.
- **Repeatability:** Complex parts can be reproduced consistently with minimal variation.
- **Efficiency:** Automation reduces machining time and labor costs.
- **Complexity:** CNC lathes can handle intricate geometries and complex toolpaths that would be difficult or impossible to achieve manually.

9.2 Live Tooling

Live tooling expands the capabilities of CNC lathes by enabling milling, drilling, and other operations without removing the workpiece from the machine. Live tools are powered by a separate motor and can be rotated at high speeds, allowing for versatile machining.

Types of Live Tools

- **Driven Tools:** These tools have their own motor and can be rotated at high speeds for milling, drilling, and tapping operations.
- **Static Tools:** These tools are not powered and rely on the lathe's spindle rotation for cutting. They are typically used for turning and boring operations.

Applications of Live Tooling

- **Milling Flats, Slots, and Keyways:** Live tools can create various features on the workpiece without requiring secondary operations on a milling machine.
- **Drilling and Tapping Holes:** Live tools can be used to drill and tap holes on the workpiece, eliminating the need for separate drilling operations.
- **Complex Geometries:** Live tooling enables the machining of complex geometries with multiple features and angles.

9.3 Multi-Axis Turning

Multi-axis turning involves the simultaneous movement of the cutting tool along multiple axes, allowing for the production of complex, contoured parts in a single setup. Common multi-axis turning configurations include:

- **Y-Axis Turning:** The tool moves along the Y-axis, perpendicular to the X and Z axes, enabling the machining of off-center features and contours.
- **B-Axis Turning:** The tool rotates around the Y-axis, allowing for complex angular cuts and the creation of features on multiple sides of the workpiece.
- **C-Axis Turning:** The spindle rotates independently of the workpiece, enabling the machining of complex contours and features on the face of the workpiece.

Benefits of Multi-Axis Turning

- **Reduced Setup Time:** Multiple operations can be performed in a single setup, eliminating the need for repositioning the workpiece.
- **Improved Accuracy:** Simultaneous movement along multiple axes increases accuracy and eliminates errors caused by repositioning.
- **Complex Geometries:** Multi-axis turning enables the production of intricate shapes and features that would be difficult or impossible to achieve with traditional turning methods.

9.4 Gear Cutting

Gear cutting on a lathe involves the creation of gear teeth on a cylindrical blank. This process requires specialized tooling and precise indexing to ensure accurate tooth spacing and profile.

Gear Cutting Methods

- **Form Cutting:** A form tool with the desired gear tooth profile is used to cut each tooth individually. This method is suitable for small batches and custom gears.
- **Generating:** A generating tool with multiple cutting edges is used to create the gear teeth by rolling the tool and workpiece together. This method is faster and more efficient for large-scale production.
- **Hobbing:** A hob, a cylindrical cutting tool with helical teeth, is used to generate the gear teeth by meshing with the workpiece blank.

This method is commonly used for high-precision gears.

Tooling for Gear Cutting

- **Form Tools:** Designed for specific gear tooth profiles and modules.
- **Generating Cutters:** Available in various types, including shaper cutters, rack cutters, and pinion cutters.
- **Hobs:** Cylindrical tools with helical teeth, designed for specific gear modules and pressure angles.

9.5 Spline Cutting

Spline cutting involves creating splines, which are keyways or grooves cut along the length of a shaft, on a lathe. Splines are used to transmit torque and rotary motion between shafts and hubs.

Spline Cutting Methods

- **Single-Point Cutting:** A single-point tool is used to cut each spline individually. This method is suitable for small batches and custom splines.
- **Broaching:** A broach, a long tool with multiple cutting teeth, is pushed or pulled through the workpiece to create the splines in a single pass. This method is faster and more efficient for mass production.
- **Spline Hobbing:** A spline hob, similar to a gear hob, is used to generate the spline teeth by meshing with the workpiece. This method is

precise and suitable for high-volume production.

Tooling for Spline Cutting

- **Single-Point Tools:** Designed for specific spline profiles and sizes.
- **Broaches:** Long tools with multiple cutting teeth, designed for specific spline profiles and sizes.
- **Spline Hobs:** Cylindrical tools with helical teeth, designed for specific spline modules and pressure angles.

9.6 Custom Tooling

Custom tooling refers to the design and fabrication of specialized tools to meet specific machining requirements. This can involve modifying existing tools or creating entirely new tools from scratch.

Benefits of Custom Tooling

- **Enhanced Productivity:** Custom tools can be designed to optimize specific operations, reducing cycle times and increasing throughput.
- **Improved Quality:** Custom tools can be tailored to achieve specific surface finishes or tolerances that may not be possible with standard tools.
- **Unique Applications:** Custom tools can be created for unique or specialized machining tasks that cannot be performed with standard tools.

Custom Tooling Design and Fabrication

The design and fabrication of custom tooling require a thorough understanding of the machining process, the workpiece material, and the desired outcome. Computer-aided design (CAD) software and CNC machining are often used to create precise and complex tool geometries.

9.7 High-Precision Machining

High-precision machining refers to the production of parts with extremely tight tolerances and exceptional surface finishes. It requires advanced techniques, specialized equipment, and meticulous attention to detail.

Factors Affecting High-Precision Machining

- **Machine Accuracy:** The lathe's precision and rigidity are crucial for achieving tight tolerances.
- **Tooling:** High-quality cutting tools and precise toolholders are essential for maintaining accuracy and minimizing tool deflection.
- **Workholding:** The workpiece must be held securely and accurately to prevent movement or vibration.
- **Environmental Factors:** Temperature variations and vibrations can affect machining accuracy.
- **Operator Skill:** The machinist's skill and experience play a significant role in achieving high-precision results.

Techniques for High-Precision Machining

- **Fine Finishing:** Using fine-grained cutting tools and light cuts to achieve a smooth surface finish.
- **Diamond Turning:** Utilizing diamond tools for ultra-precise turning of non-ferrous materials.
- **Grinding:** A precision abrasive machining process used to remove material and achieve tight tolerances and smooth surface finishes.

By exploring these advanced lathe operations, you can push the boundaries of your machining capabilities and produce high-quality, complex parts with exceptional precision. The combination of CNC programming, live tooling, multi-axis turning, gear and spline cutting, custom tooling, and high-precision machining techniques opens up a world of possibilities for creative and innovative manufacturing.

CHAPTER 10

MAINTENANCE, SAFETY, AND TROUBLESHOOTING

Ensuring the longevity, optimal performance, and safety of your metal lathe requires a combination of regular maintenance, proactive troubleshooting, and adherence to advanced safety protocols. This chapter delves into these crucial aspects, empowering you to maintain your lathe in peak condition, address common issues effectively, and create a safe and enjoyable machining environment.

10.1 Regular Maintenance Tasks

A well-maintained lathe is a reliable and productive tool. By adhering to a regular maintenance schedule, you can prevent unexpected breakdowns, extend the lifespan of your machine, and ensure consistent accuracy in your work.

Daily Maintenance Tasks

- **Cleaning:** After each use, thoroughly clean the lathe bed, carriage, cross slide, and other components to remove chips, debris, and coolant residue. Use a brush, compressed air, or a shop vacuum to ensure a clean working surface.
- **Lubrication:** Regularly lubricate the ways, lead screw, cross slide screw, and other moving parts according to the manufacturer's recommendations. Use the appropriate type of

lubricant (e.g., way oil, grease) and apply it sparingly to avoid excess buildup.
- **Tool Inspection:** Inspect cutting tools for wear, damage, or dullness. Sharpen or replace tools as needed to maintain optimal cutting performance and prevent damage to the workpiece.
- **Chuck Inspection:** Check the chuck jaws for proper alignment, tightness, and damage. Clean and lubricate the chuck as needed.
- **Tailstock Alignment:** Verify the tailstock alignment to ensure accurate turning between centers. Adjust the tailstock if necessary to maintain parallelism with the lathe bed.

Weekly Maintenance Tasks

- **Way cleaning and Lubrication:** Perform a more thorough cleaning and lubrication of the ways, paying attention to areas that may be prone to wear.
- **Headstock Lubrication:** Lubricate the headstock bearings and gears according to the manufacturer's instructions. This may involve adding oil to oil cups or grease to grease fittings.
- **Motor Inspection:** Check the motor for proper operation and lubrication. Ensure the motor is clean and free of debris, and check for any unusual noises or vibrations.
- **Electrical Connections:** Inspect all electrical connections for tightness and corrosion. Clean or replace any corroded connections.

Monthly Maintenance Tasks

- **Leveling:** Verify the lathe's levelness using a precision level. Adjust the leveling feet as needed to maintain accuracy and prevent uneven wear on components.
- **Gib Adjustment:** Check the gibs on the carriage and cross slide for proper tightness. Adjust the gib screws to eliminate excessive play or tightness, ensuring smooth and controlled movement.
- **Belt Inspection:** Inspect the belts for wear, tension, and alignment. Adjust or replace belts as needed to maintain proper power transmission and prevent slippage.
- **Coolant System:** Clean the coolant system reservoir and filters. Replace the coolant if it becomes contaminated or loses its effectiveness.

Annual Maintenance Tasks

- **Professional Inspection:** Have a qualified technician perform a comprehensive inspection and maintenance of your lathe. This may include checking for wear on critical components, replacing worn parts, and adjusting backlash in gears and screws.

10.2 Troubleshooting Common Issues

Even with regular maintenance, lathes can experience issues that affect their performance and accuracy. Recognizing and addressing these problems promptly can save time, prevent further damage, and ensure smooth operation.

Common Lathe Issues and Solutions

- **Chatter:**
    - **Causes:** Loose tool holders, worn bearings, incorrect tool geometry, excessive overhang, or incorrect cutting parameters.
    - **Solutions:** Tighten tool holders, replace worn bearings, correct tool geometry, reduce overhang, or adjust cutting parameters (speed, feed, depth of cut).
- **Poor Surface Finish:**
    - **Causes:** Dull tools, incorrect cutting speed or feed, inadequate lubrication, or vibration.
    - **Solutions:** Sharpen or replace tools, adjust cutting parameters, apply appropriate cutting fluid, or identify and eliminate sources of vibration.
- **Taper:**
    - **Causes:** Misaligned tailstock, uneven wear on the ways, or incorrect tool setting.
    - **Solutions:** Realign the tailstock, rescrape or replace worn ways, or correct the tool height.
- **Overheating:**
    - **Causes:** Excessive cutting speed, insufficient lubrication, dull tools, or inadequate chip removal.
    - **Solutions:** Reduce cutting speed, increase lubrication or coolant flow,

sharpen or replace tools, or improve chip removal.

- **Tool Breakage:**
    - **Causes:** Excessive cutting forces, incorrect tool geometry, incorrect tool material, or inadequate clamping.
    - **Solutions:** Reduce cutting forces, correct tool geometry, use a more suitable tool material, or improve tool clamping.

Troubleshooting Tips

- **Isolate the Problem:** Determine the specific issue you're facing by carefully observing the lathe's behavior and the workpiece's condition.
- **Consult the Manual:** Refer to the lathe's user manual for troubleshooting guidelines and recommended solutions.
- **Seek Expert Advice:** If you're unsure how to resolve a problem, seek help from a qualified technician or experienced machinist.
- **Record Maintenance and Repairs:** Keep a log of maintenance tasks performed and any repairs made to track the lathe's history and identify recurring issues.

10.3 Advanced Safety Protocols

While basic safety precautions are essential for lathe operation, advanced safety protocols provide an additional layer of protection, especially when working with complex setups, large workpieces, or hazardous materials.

Advanced Safety Protocols

- **Risk Assessment:** Before starting any machining operation, conduct a thorough risk assessment to identify potential hazards and implement appropriate control measures.
- **Machine Guarding:** Install and use machine guards to protect operators from flying chips, rotating parts, and other hazards.
- **Emergency Stop:** Ensure the emergency stop button is easily accessible and functioning properly.
- **Lockout/Tagout:** Implement lockout/tagout procedures to prevent accidental machine startup during maintenance or repairs.
- **Personal Protective Equipment (PPE):** In addition to safety glasses and hearing protection, consider using face shields, respirators, or other specialized PPE when working with hazardous materials or processes.
- **Training and Certification:** Ensure all operators are properly trained and certified to operate the lathe and perform specific tasks.
- **Regular Safety Inspections:** Conduct regular safety inspections of the lathe and surrounding work area to identify and address potential hazards.

10.4 Lathe Upgrades and Modifications

Upgrading or modifying your lathe can enhance its capabilities, improve its performance, and extend its lifespan. However, it's crucial to carefully consider the potential benefits and risks before making any modifications.

Common Lathe Upgrades

- **Quick Change Tool Post:** This upgrade allows for faster and easier tool changes, reducing setup time and increasing productivity.
- **Digital Readouts (DROs):** DROs provide precise, real-time feedback on tool position and movement, improving accuracy and reducing the need for manual measurement.
- **Variable Speed Drives (VSDs):** VSDs allow for precise control of spindle speed, enabling optimization for different materials and operations.
- **Coolant Systems:** Upgrading to a more efficient or higher-capacity coolant system can improve chip evacuation, tool life, and surface finish.
- **CNC Conversion:** Converting a manual lathe to CNC operation can significantly increase its capabilities, automation, and precision.

Considerations for Modifications

- **Compatibility:** Ensure any upgrades or modifications are compatible with your lathe model and manufacturer's specifications.
- **Safety:** Prioritize safety when making modifications. Follow proper procedures and consult with experts if needed.
- **Warranty:** Be aware that unauthorized modifications may void your lathe's warranty.

- **Cost-Benefit Analysis:** Weigh the costs of upgrades and modifications against their potential benefits to determine if they are worthwhile investments.

10.5 Resources for Further Learning

Continuous learning is essential for improving your lathe skills and staying up-to-date with the latest techniques and technologies. Numerous resources are available to help you expand your knowledge and expertise:

- **Books and Manuals:** Technical books and manuals on lathe operation, tooling, and projects can provide valuable information and guidance.
- **Online Courses and Tutorials:** Many online platforms offer courses and tutorials on lathe operation, ranging from beginner to advanced levels.
- **Machinist Forums and Communities:** Online forums and communities provide a platform for connecting with other machinists, exchanging knowledge, and seeking advice.
- **Workshops and Seminars:** Attend workshops and seminars offered by experienced machinists or training centers to learn new techniques and best practices.
- **Manufacturer Resources:** Consult the lathe manufacturer's website or customer support for manuals, troubleshooting guides, and other resources.

10.6 Building a Lathe Community

Connecting with other lathe enthusiasts can be a valuable source of inspiration, support, and knowledge sharing. There are several ways to build a lathe community:

- **Online Forums and Groups:** Join online forums and social media groups dedicated to metalworking and lathe operation. Participate in discussions, share your projects, and ask questions.
- **Local Clubs and Organizations:** Look for local machinist clubs or organizations where you can meet fellow enthusiasts, attend workshops, and share ideas.
- **Mentoring:** Seek out experienced machinists who are willing to share their knowledge and expertise.
- **Collaborate on Projects:** Collaborate with other machinists on joint projects to learn new skills and tackle challenging tasks.

10.7 Sharing Your Lathe Creations

Sharing your lathe projects with others is a great way to showcase your skills, inspire others, and contribute to the broader machining community. Several platforms and approaches can facilitate this process.

Online Platforms for Sharing

- **Social Media:** Platforms like Instagram, Facebook, and Pinterest are excellent for showcasing photos and videos of your projects. Use relevant hashtags (e.g., #metalworking, #latheprojects) to reach a wider audience.
- **YouTube:** Create video tutorials or project demonstrations to share your knowledge and techniques with others.
- **Maker Forums and Communities:** Participate in online forums and communities dedicated to machining and craftsmanship. Share your projects, ask questions, and engage in discussions with fellow enthusiasts.
- **Personal Blogs or Websites:** Consider creating a personal blog or website to document your projects, share your experiences, and provide tutorials or resources for others.

Other Ways to Share Your Creations

- **Local Exhibitions and Fairs:** Participate in local exhibitions and fairs showcasing handmade crafts and artisan works.
- **Gifts and Donations:** Share your creations with friends and family as unique gifts or donate them to charity auctions or events.
- **Teach and Mentor:** Offer workshops or mentoring to aspiring machinists to share your knowledge and passion for the craft.

By actively sharing your lathe projects, you contribute to a vibrant and supportive community of makers. You can inspire others to explore their creativity, learn new skills, and discover the joy of working with their hands.

Conclusion

Mastering lathe operations and exploring advanced techniques is an ongoing journey of learning and discovery. This chapter has equipped you with the knowledge and tools to maintain your lathe, troubleshoot common issues, prioritize safety, explore advanced techniques, and become an active member of the machining community. Embrace the challenges and rewards of lathe work, and continue to hone your skills as you embark on increasingly complex and fulfilling projects.

Dear reader, Thanks for reading!!!

www.ingramcontent.com/pod-product-compliance
Lightning Source LLC
Chambersburg PA
CBHW050233230526
45470CB00005B/1926